Web Wizards

PATHFINDER EDITION

By Rebecca L. Johnson

CONTENTS

2 Web Wizards
8 A Spinning Sampler
10 Veggie Spider!
12 Concept Check

Keeping Watch. A female fishing spider guards a round, white sac. Inside the sac are many tiny spider eggs.

Web Wizards

Spiders are everywhere. You can find them in dusty attics and damp basements. You can spot spiders in forests and fields. Earth is home to about 40,000 different kinds. They range from tiny jumping spiders to bird-eating spiders as big as your hand. Let's find out more about these amazing creatures.

By Rebecca L. Johnson

What Are Spiders?

Many people think spiders are insects. But a spider is an **arachnid**. How can you tell the difference? A spider has eight legs, while an insect has six. A spider's body has a head and an **abdomen**. That's its large back part. An insect's body has a head and an abdomen, too. But it also has a third part in between, called a thorax.

Spider Parts. You can see the spider's eight legs and two body parts. How is a spider's body different from a fly's body?

Super Silk

Spiders can make threadlike strands called silk. Silk comes out of special structures on a spider's abdomen. Each structure looks like a nozzle at the end of a hose. These silk-making structures are spinnerets.

Silk comes shooting out of a spider's **spinnerets**. It is both stretchy and strong. A strand of spider silk is about five times stronger than a steel wire with the same thickness. Spiders can produce as many as seven kinds of silk. They use different kinds for different jobs.

Spinning Silk. This Mediterranean black widow spider shoots silk out of its abdomen to spin a web.

Tasty Treat. The wasp spider stings its prey before eating it. The wasp spider's venom is poisonous to insects, but not people.

Wonderful Webs

Nearly all spiders are hunters. They need to catch and eat insects or other small animals for food. Many spiders use silk to make webs. Most webs are traps for insects and other small animals that spiders eat. Spiders are good at making webs that are hard to see. You know that's true if you've ever bumped into one!

Webs have some silk strands that are strong and smooth. But other strands are as sticky as glue. When insects or small animals fly into webs, they get stuck on these sticky strands.

Ropes and Wraps

Have you ever seen a spider hanging by a silk strand? Spiders also use silklike ropes. They use these silk ropes to get down from high places, almost like mountain climbers. Jumping spiders attach a silk strand to a surface before they leap. If they miss their target, this safety rope keeps them from crashing.

Female spiders use a special type of silk to wrap up their eggs. They make an egg sac in which their eggs will be safe. An egg sac can contain dozens, or even hundreds, of eggs.

Tightrope Walker. This jumping spider jumps from one plant to another using silk strands.

A Spider Life Cycle

The stages in the life of a plant or animal make up its life cycle. A spider's **life cycle** begins with an egg inside an egg sac. Some spider mothers stay with their egg sac until the eggs hatch. Others hide the sac in a safe place and then leave.

1

Spider eggs usually hatch in a few weeks. A baby spider, or spiderling, comes out of each one. A spiderling is tiny. But it looks just like its parents. That's because the mother and father passed on their features to the spiderling. These features are called **inherited traits**. Eye color, abdomen shape, and leg length are examples of inherited traits. Something a spider does can be an inherited trait, too.

5

After their last molt, the you
spiders are adults. Each fem
spider will go on to lay her
eggs. She will wrap them in
made of silk. The eggs will
and the spider life cycle will
be complete.

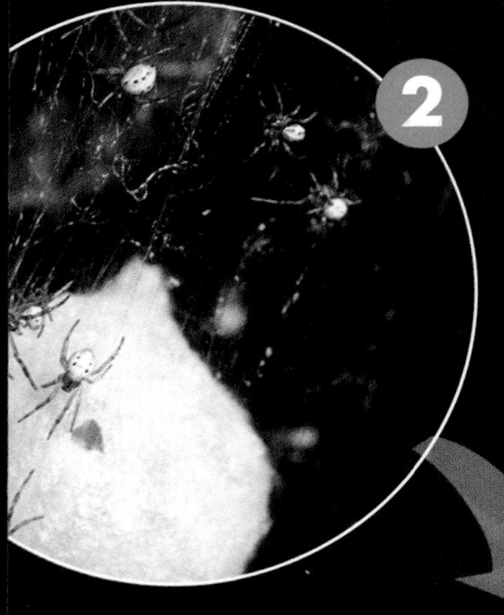

Spiders have a hard body covering. It doesn't stretch. In order to grow, spiders must **molt**. This means they replace their body covering with a newer, bigger one. The new body covering forms under the old one. Then the old one splits open and a new, slightly bigger covering pushes through. Spiderlings usually molt once in their egg sac. Then they make a hole in the sac and crawl out.

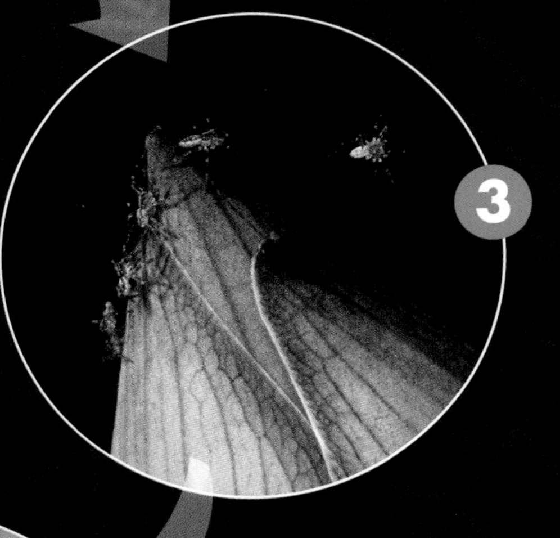

The spiderlings head off on their own. Some kinds simply crawl away. Others make strands of silk that catch the breeze. The baby spiders float away as though they were tied to balloons or kites. The wind carries them to new homes.

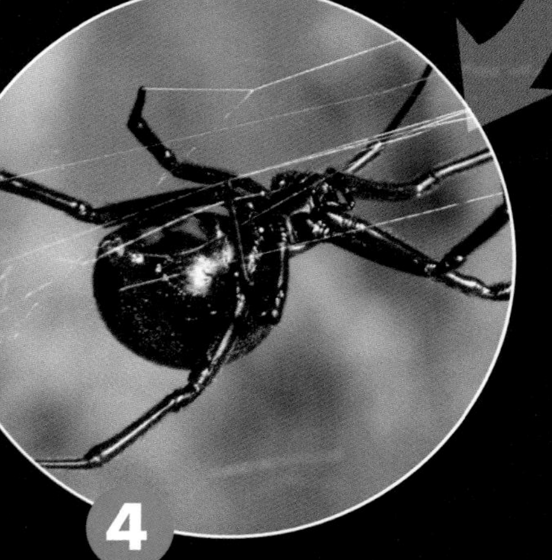

As the weeks pass, the spiderlings molt several more times. Each time they are bigger. They spin their own webs to catch food. Their webs look just like the webs their parents made. That's because web-making is an inherited trait.

Wordwise

abdomen: the rear part of a spider's body

arachnid: an invertebrate animal with a hard body covering and eight legs

inherited trait: a feature or behavior passed down from parents

life cycle: all the different stages in the life of a living thing

molt: to get rid of an old body covering and replace it with a new, slightly larger one

spinneret: spider body part that makes silk

A Spinning Sampler

Different kinds of spiders make different kinds of webs. They twist and loop and weave silk strands with the tips of their legs. Web-making spiders work quickly. They are careful not to get caught on their own sticky strands!

Sheet Web

Sheet webs are flat sheets of silk strands. A spider that builds this kind of web hangs upside down beneath it. When an insect lands on top of the sheet, the spider moves fast. It grabs the insect and pulls it down through the sheet.

Orb Web

An orb web is very organized. It looks like a wheel with spokes. It also has sticky strands that circle around and around from the center to the edge. The spider often sits right in the middle of its web.

Tangle Web

Tangle webs look very messy. The silk strands go every which way. A good place to find tangle webs is in corners near the ceiling.

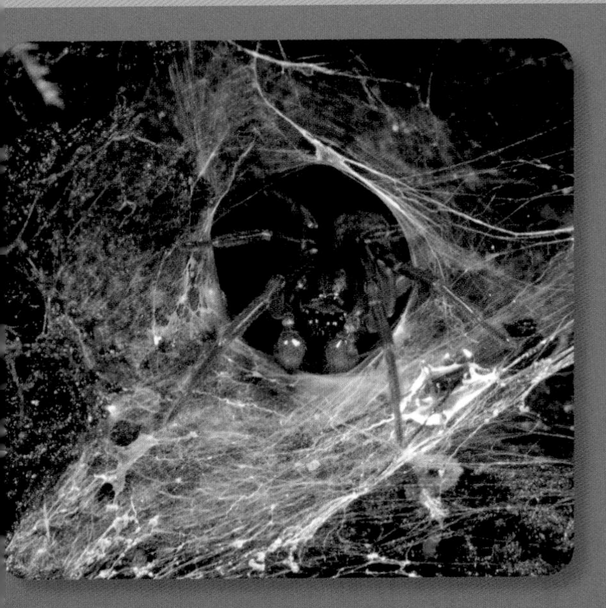

Funnel Web

Not surprisingly, funnel webs are shaped like a funnel. They are wide at one end and narrow at the other. The spider hides in the narrow part. Insects land on the wide part and get caught on sticky strands. The spider dashes out and grabs them.

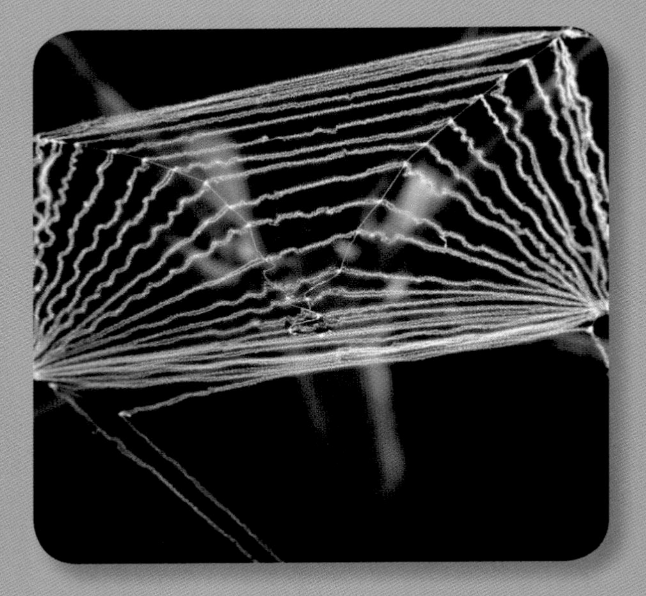

The Net

Net-casting spiders have long legs and very good eyesight. They spin webs that they hold between their legs. The spider waits for insects to fly by. Then the spider tosses its net and catches them.

Veggie Spider!

For a long time, people thought all spiders ate other animals for food. Recently, scientists working in Central America got a surprise. They discovered a small jumping spider that eats plants. It's the first veggie spider known to science.

The spider eats the tiny sweet tips of leaves that grow on a certain kind of plant. Ants eat the sweet leaf tips, too. In fact, the ants guard the leaves. They will attack anything that tries to steal their favorite food.

The spider watches the ants from a safe spot. When the ants aren't paying attention, it runs toward a leaf tip. If the ants move in, the spider jumps out of reach. Or it may jump off the leaf and hang by a strand of silk. There, it is safely out of reach. But the veggie spider is a speedy spider. It is much faster than the ants. It gets away with a sweet leaf tip nearly every time.

Standing Guard. This acacia ant is guarding the leaf tips on an acacia plant. Will the veggie spider be able to outsmart the ant to get the leaf?

Web Wizards

Think you're a wizard when it comes to spiders? Prove it by answering the questions below.

1. How can you tell the difference between a spider and an insect?
2. What are spinnerets?
3. Name three different ways that a spider can use silk.
4. How do young spiders know what kind of web to build?
5. Why is the veggie spider so special?